MY SENSES

Sight

CHRISTINA EARLEY

A Crabtree Roots Book

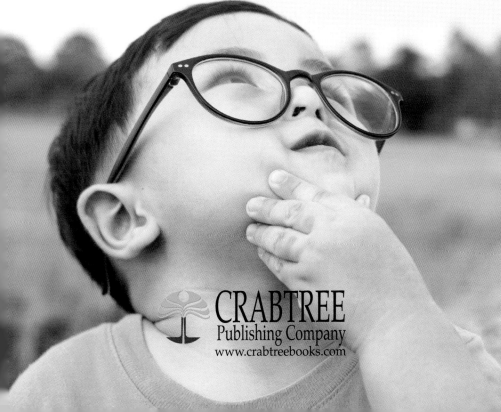

CRABTREE
Publishing Company
www.crabtreebooks.com

School-to-Home Support for Caregivers and Teachers

This book helps children grow by letting them practice reading. Here are a few guiding questions to help the reader with building his or her comprehension skills. Possible answers appear here in red.

Before Reading:

- What do I think this book is about?
    - *I think this book is about my sense of sight.*
    - *I think this book is about how I use my eyes.*

- What do I want to learn about this topic?
    - *I want to learn about the part of my body I use for sight.*
    - *I want to learn how my eyes keep me safe.*

During Reading:

- I wonder why…
    - *I wonder why my eyes can see many colors.*
    - *I wonder why my eyes can see many different things.*

- What have I learned so far?
    - *I have learned that I use my eyes to see animals.*
    - *I have learned my eyes can see small and large things.*

After Reading:

- What details did I learn about this topic?
    - *I have learned that sight is one of my five senses.*
    - *I have learned I can use my eyes to read.*

- Read the book again and look for the vocabulary words.
    - *I see the word **eyes** on page 4 and the word **safe** on page 8. The other vocabulary words are found on page 14.*

Sight is one of my five **senses**.

I use my **eyes** to see.

I see pretty **colors**.

Look at the
many animals.

My eyes help keep me **safe**.

See what is **small**.

See what is **large**.

I use my sight to learn.

Word List

Sight Words

animals	is	me	the
at	keep	my	to
five	learn	of	use
help	look	one	what
I	many	see	

Words to Know

colors

eyes

large

safe

senses

small

42 Words

Sight is one of my five **senses**.

I use my **eyes** to see.

I see pretty **colors**.

Look at the many animals.

My eyes help keep me **safe**.

See what is **small**.

See what is **large**.

I use my sight to learn.

Written by: Christina Earley
Designed by: Rhea Wallace
Series Development: James Earley
Proofreader: Janine Deschenes
Educational Consultant: Marie Lemke M.Ed.

Photographs:
Shutterstock: Chinnapong: cover; Oekka.k: p.1; wavebreakmedia: p. 3, 13, 14; Mariia Khamidulina: p. 4, 5, 14; Jorra: p. 6; Serget Novikov: p. 9, 14; IM_photo: p. 11, 14

Library and Archives Canada Cataloguing in Publication

Available at the Library and Archives Canada

Library of Congress Cataloging-in-Publication Data

Available at the Library of Congress

Crabtree Publishing Company
www.crabtreebooks.com   1-800-387-7650

Printed in the U.S.A./062021/CG20210401

Copyright © 2022 **CRABTREE PUBLISHING COMPANY**

All rights reserved. No part of this publication may be reproduced, stored in a retrieval system or be transmitted in any form or by any means, electronic, mechanical, photocopying, recording, or otherwise, without the prior written permission of Crabtree Publishing Company. In Canada: We acknowledge the financial support of the Government of Canada through the Canada Book Fund for our publishing activities.

Published in the United States
Crabtree Publishing
347 Fifth Avenue, Suite 1402-145
New York, NY, 10016

Published in Canada
Crabtree Publishing
616 Welland Ave.
St. Catharines, Ontario L2M 5V6